YOUR KNOWLEDGE HAS VALUE

- We will publish your bachelor's and
  master's thesis, essays and papers

- Your own eBook and book -
  sold worldwide in all relevant shops

- Earn money with each sale

Upload your text at www.GRIN.com
and publish for free

Bibliographic information published by the German National Library:

The German National Library lists this publication in the National Bibliography;
detailed bibliographic data are available on the Internet at http://dnb.dnb.de .

Imprint:

Copyright © 2018 GRIN Verlag
Print and binding: Books on Demand GmbH, Norderstedt Germany
ISBN: 9783668811638

This book at GRIN:

https://www.grin.com/document/444210

Adongoi Toakodi

The Impact of Sea Robbery on Artisanal Fishing. The Niger Delta Experience

GRIN Verlag

The Impact of Sea Robbery on Artisanal Fishing: The Niger Delta Experience

Adongoi, Toakodi (Ph.D)
Ogbia Rural Development Authority,
Opume Town, Ogbia Local Government Area,
Bayelsa State Nigeria

Contents

ABSTRACT

The study examined the impact of sea robbery on artisanal fishing in Niger Delta region of Nigeria. The study is survey based; A Sample of 400 was derived using Taro Yamane sample size determination technique. Questionnaires and oral interview were the instruments used for data collection. *The data collected were analyzed using Pearson Product Moment Correlation Coefficient (PPMC). The results revealed that there is a significant relationship between sea robbery and artisanal fishing which implies that the continuing existence of the activities of sea robbers in our territorial waters will result to poor fish harvest because fishermen are scared of fishing in deep waters where there is fish abundance. Consequent upon this, it is recommended, among others, community policing especially hotspots communities along the waterways and creeks of the Niger Delta region.*

Key words: Impact, sea robbery, artisanal fishing, Niger Delta Experience.

Introduction

Fishing is an important occupation of the people of Nigeria and the Niger Delta region in particular. Next to the oil and Gas industry, the fishing industry is a huge foreign exchange earner for the country. Besides this, fishing also provides a great part of the protein needs of the country (Jamabo and Ibim, 2010; Tawari and Davies, 2010). The fishing industry in Nigeria also provides employment and engenders economic growth (Pittman, 2011). However, the activities of the multinational oil companies have so polluted the environment that the fishing industry is in a state of decline (Nurudeen, 2015). As if this is not a bad enough blow to the industry, the activities of sea robbers have added a new dimension to the difficulties confronting fishing business in Nigeria especially in the Niger Delta region. Violent attacks by sea robbers against the fishing industry in Nigeria have caused the fleet of fishing trawlers to dwindle (Usim, 2016). The immediate effect of this situation has been loss of income and unemployment (Terzi, 2012; Wajilda, 2013; Ezem, 2012, 2013).

The activities of sea robbers have impacted negatively on the economy of the Niger Delta region. Evidently, the actions and inactions of sea robbers have adversely affected fishing business as well as other commercial activities in the region. There is a reduction in the fleet of Nigerian fishing trawlers on these waters leading to a huge loss of revenue to the affected fishermen and traders. Both the Federal and the State Governments have instituted policies to tackle the menace of sea robbery in the region, but these measures such as "Operation Restore Hope, Operation Pulo Shield, Operation Delta Safe and, recently, Operation Crocodile Smile", have not achieved the expected results. Such measures by security agents such as the Navy, Maritime Police, Nigeria Security and Civil Defence Corp, the Joint Task Force (JTF) all in a bid to curb sea robbery in the region have not succeeded in eliminating the problem.

A number of social science scholars have studied the phenomenon of sea robbery in Nigeria (Onuoha and Hassan, 2009; Phelps, 2010; Nicoll, 2012; Dogarawa, 2013; Otto, 2015).These studies on sea robbery appear to have focused more on its effects on large commercial boats (shipping industry), industrial fishing (trawling) as well as the Oil and Gas industry. Although small-scale fishing industry and the likes on the waterways are the major areas of attacks by sea robbers, existing studies seem to have ignored this group of victims. Similarly, available literature on sea robbery and national security such as Onuoha and Hassan (2009), Udensi, Okpara and Oyinyechi (2014), etc, have not established the link between maritime insecurity and the decline in small-scale maritime business operations in the study area. There is a dearth of literature, particularly empirical one, on the relationship(s) between sea robbery and depletion of artisanal fishing generally, and the Niger Delta region in particular. This study, therefore, has focused on sea robbery and its effect on small-scale fishing, which earlier studies seem to have ignored. Therefore, this study fills the gap in the existing literature on the subject matter in the littoral states of Niger Delta.

In Nigeria, sea robbery and other types of maritime criminal behaviour have not only negatively affected the oil and gas industry but have also affected local fisheries (Mpi, 2011; Ochai, 2013; Ships and Ports, 2014) and international trade (Sanga, 2010). Fishing constitutes one of Nigeria's most significant non-hydrocarbon exports, yet sea robbery and other types of violence have devastated the sector (Perouse de Montclos, 2012). Fishing is the second highest non-oil export industry in Nigeria, and pirate attacks on fishing trawlers have reached the point that many fishing boat captains refuse to sail. Nigeria stands to lose up to US$600 million in export earnings due to piracy threats to its fisheries (George, 2015).

There are no good statistics on the number of attacks on fishermen in Nigeria territorial waters, but newspaper articles and radio commentaries show that hundreds of attacks are launched on fishermen each year. Consequently, it appears that fishermen are relatively suffering the highest economic losses because of sea robbery and illegal, unreported and unregulated (IUU) fishing attacks. Both artisanal and industrial fishermen are often the victims of sea robbery attacks (Ships and Ports, 2014; Zircon Marine Ltd., 2014). Attacks on trawlers show that this industry loses millions of naira each year and many fishermen are killed during attacks. Attacks on fishermen involve stealing their catch, engine, fuel, personal belongings and, at times, even their vessels (Graf, 2011; International Peace Institute, 2014; Zircon Marine Ltd., 2014). Sea robbers also use fishermen as human shield or disguise during their attacks on more profitable targets. Fishermen already suffer the consequences of overexploitation of fish and often cannot afford to replace their stolen items. This has forced many fishermen to give up their occupation, and due to their limited employment opportunities, this has forced some of them to engage in illegal activities, including sea robbery (Orji, 2013).

It is estimated that Nigeria loses about 26.3 billion US Dollars annually to various criminality including piracy and sea robbery (Oyetunji, 2012). Specifically, Ezem (2012) in his report quoted Mr. Joseph Overo (Then President of the Nigeria Trawlers Association) to have raised alarm over the menace of sea robbers, saying "the industrial fishing sub-sector in Nigeria lost in excess of 119 billion Naira in the last eight years in fishing revenue alone". In his opening remarks during a workshop on "Harnessing the Potentials of Nigeria's Maritime Sector for Sustainable Economic Development", former President of the Federal Republic of Nigeria, President Goodluck Jonathan (represented by the then Minister of Finance and Coordinating Minister for the economy, Dr. Ngozi Okonjo -Iweala), lamented that piracy in the Gulf of Guinea has threatened about 600 million US Dollars worth of fishing exports. According to him, the cost of piracy to our economy is unacceptably high. Pirates frustrate fishing activities and threaten investments prospect in the West African Coast (Oyetunji, 2012).

Onuoha and Hassan (2009) have also noted the viability of fishing business to the nation's economy. Maritime trade is a significant contributor to Nigeria's economic development especially in the area of fishing business. Ochai (2013) further observed that the price of sea-foods is likely to increase soon if steps are not taken to check the increasing rate of sea pirates on the Nigerian waters.

In a report made available by Leadership Newspaper of 26 February, 2016, scores of indigenes from the coastal communities of Middleton, Koluama, Fish town, Akassa, Pennington and Brass in Bayelsa State of Nigeria threatened to engage foreign fishing trawler owners in a bloody clash over alleged armed attacks on local fishermen in the area. The local fishermen from the aggrieved communities in protest letters to the leadership of the Ijaw Youths Council (IYC) and other security agencies accused the foreign fishing trawler owners operating on the high sea of indiscriminate gun attacks on local fishermen engaged in fishing activities close to their trawlers. While the leaders of the aggrieved communities claimed that the gun attacks had led to some fishermen sustaining serious injuries and thus being unable to fish on the high sea, some concerned indigenes excused the decision by the foreign fishing operators to procure arms and defend themselves, claiming that the decision was based on increased armed attacks from sea robbers and unknown criminal elements from the host communities (Okhomina, 2016). It was learnt that some of the aggrieved communities accused the Chinese fishing trawler owners of the armed attacks on the indigenes of the communities. But a source close to some of the foreign operators informed LEADERSHIP newspaper that the aggrieved communities have, in the last few months, become hotspots for sea robbers to show their abilities on defenseless seafarers by way of forcefully collecting their valuables (Okhomina, 2016).

Usim (2016) reported that Nigeria's multi-million dollar fish trawling business is in danger of total extinction as incessant sea robbers' attacks have continued to scare away trawler owners and seafarers from their lucrative business. This situation has resulted in maritime stakeholders lament that the nation's territorial waters are now dangerous for seafarers on commercial ferry boats, fish trawlers and other crafts for fear of losing their consignment and lives in most cases. Usim (2014) further reiterated that the patrol operations by the Nigerian Maritime Administration and Safety Agency (NIMASA) and the Nigerian Navy are grossly inadequate and unsatisfactory given that much of Nigeria's territorial waters are poorly policed. Only in January 2016 as reported by Usim (2016), dare-devil robbers abducted two out of 14 crew members – the Captain and Chief Engineer, aboard a trawler vessel, MV KULAK IX, off Dodo River in Bayelsa State, Nigeria.

Despite the fact that much of the cargo aboard the vessel belonging to Barnaly Fisheries Nigeria Limited were not stolen by the sea robbers, the incident frightened other seafarers fishing along the nation's waters as they reportedly fled the sea even without any catch. Economic experts say fish-trawling business employs thousands of Nigerian youths and contributed significantly to the nation's economy between the early 1980's up till early 2000's when sea robbers' attacks were much less (Aderigbola, 2015). But local trawler operators are alleging that foreigners who are eyeing the business are now sponsoring mercenaries to attack them with a view to taking over the business (Usim, 2016).

On the effect of piracy on fish trawling business, Margaret Orakwusi, a former President of the Nigeria Trawlers Association (NITOA), lamented that continued robbers' attacks has forced a lot of indigenous operators out of the business. She further noted that the number of companies operating in the sector in 2005 and 2006 was about 39, but this has drastically reduced to nine by 2014 (Usim, 2016). Ojo (2016) in a study stated that fish trawling is a capital-intensive project and it also brings the much-needed foreign exchange.

There was a time the industry ranked second to the oil industry in foreign exchange earnings. In 2005/2006, there were about 250 trawlers, but in year 2014, it had depleted to 124. During its boom, there were about 35 companies operating in the sector but over the years, it has reduced to just nine. Regarding sea robbery attacks, it is noteworthy that most of the attacks are not being reported, probably out of frustration by the owners of the vessels who get discouraged when the reported cases fail to yield any positive results or bring succour (in financial terms) to the victims (Ojo, 2016).

Maritime piracy and sea robbery also imposes significant costs on local fishing economies. According to IMO, pirates attacked tuna vessels at least three times in 2009 as they fished 650 to 800 kilometers beyond Somali territorial waters. One vessel was captured, leading to a ransom payment that exceeded US$1 million. The threat of pirate attacks has prompted many vessels to avoid some of the richest fishing spots in the Indian Ocean. Dwindling catches have raised concern that the Seychelles and Mauritius could face severe economic problems (James, 2013).

PIRACY AND ARMED ROBBERY: A CONCEPTUAL CLARIFICATION

The ICC International Maritime Bureau (2017), defines piracy in Article 101 of the 1982 United Nations Convention on the Law of the Sea (UNCLOS) and the International Maritime Organization (IMO) in its 26th Assembly session define armed robbery as is seen in Resolution A.1025 (26).

Article 101 of UNCLOS defines Piracy as:

(a)   any illegal acts of violence or detention , or any act of devastation, committed for private ends by the crew or the passengers of a private ship or a private aircraft, and directed -

    (i)   on the high seas, against another ship or aircraft, or against persons or property on board such ship or aircraft;

(ii)   against a ship, aircraft, persons or property in a place outside the jurisdiction of any State;

(b)   any act of voluntary participation in the operation of a ship or of an aircraft with knowledge of facts making it a pirate ship or aircraft;

(c)   any act of inciting or of intentionally facilitating an act described in subparagraph (a) or (b)

The IMO on the other hand defines Armed Robbery in Resolution A.1025 (26) "Code of Practice for the Investigation of Crimes of Piracy and Armed Robbery against Ships" as any of the following acts:

1.   any illegal act of hostility or detention or any act of depredation or threat thereof, other than an act of piracy; committed for private ends and directed against a ship or against persons or property on board such a ship, within a State's internal waters, archipelagic waters and territorial sea;

2.   any act of inciting or intentionally facilitating an act described above.

This study therefore adopts IMO Resolution A.1025 (26) and defines any illegal act of violence or detention or any act of depredation or threat thereof, other than an act of piracy; committed for private ends and directed against a ship or against persons or property on board such a ship, within Nigeria's territorial waters as Sea Robbery.

Theoretical Framework

The Routine Activity Theory propounded in 1979 by Cohen and Felson in their work entitled "Social Change and Crime Rate Trends" is adopted as the theoretical framework for this study. In the "Routine Activity Approach", Cohen and Felson proposed that crime is the aftermath of combined result of three indispensable elements: First there must be a motivated offender who is competent of committing felony. Second it is not sufficient for the possible offender to be motivated; he must also be capable to execute his criminal intention.

According to Cullen and Agnew (2006), the routine activity approach is based on two rather simple ideas. First, for crime to occur, motivated offenders must meet with suitable targets in the absence of qualified guardians. Secondly, they noted that the likelihood of this situation occurring is influenced by their routine activities including the work, family, leisure, and consumption activities. For example, if we spend more time in public places such as bars and on the street, we increase the tendency that we will come into contact with motivated offenders in the absence of competent guardians. However, routine activity theory does not explain why an offender is motivated to commit a crime, but instead assumes that motivation is constant (Cohen and Felson, 1979; Morrow, 2015; Wikstromolof, 2009; United States Legal Incorporation, 2015).

Igbo (2008) noted that for a crime to occur, a motivated offender must also identify and engage a suitable target. Suitable targets can take a number of forms depending on the nature of the crime (i.e. the particular intent of the offender) and the situational context (i.e. the available opportunities). A suitable target might be an object, such as a piece of valuable property to steal or a home to burglarise.

The final element of routine activities theory consists of proficient guardianship, which bears the potential to dissuade or avert crime even in the presence of a motivated offender with a selected suitable target. Capable guardianship is an expansive concept that researchers interpret and study in a variety of ways. Formal types of guardianship such as police officers and other types of law enforcement agents, symbolise a well-recognised form of protection from crime and victimisation. Routine activities theory suggests that the existence of these agents might avert a crime from happening. Many potential offenders, despite being motivated to commit a crime, would be hesitant to engage in criminal behaviour with a police officer's presence.

The routine activity theory is relevant to this study on the impact of Sea Robbery on Artisanal Fishing: the Niger Delta Experience, because it assists to explain the existence of the crime in the littoral states of the Niger Delta Region. First, unemployed youths in the region are a pool of persons who are ready and capable of committing crime of the nature of sea robbery. Vulnerable targets are in the form of international and domestic tourists, oil tankers, fishing crafts, trawlers, speedboat operators, passengers, and local businessmen and women that ply the waterways. Most of these targets are not always well guarded. The numerous mazes of creeks in the littoral states of the Niger Delta provide hiding places for suspects or offenders who usually lay ambush for their targets in waterways that are not well protected. The absence of protection for these targets exposes the latter to incessant attacks by these motivated offenders. Besides, when faced with threat to life, such target usually panic enough to promise their assailants instant wealth reward. Such offers are quite appealing to sea robber and fuel their appetite for maritime criminality. Therefore, the routine activity theory is very useful for concisely explicating the impact of sea robbery on artisanal fishing: the Niger Delta experience.

Methodology

The research design adopted for this study was survey. It was chosen due to the explorative nature of the study. The population of the study comprised maritime business operators in the region. A multi-stage sampling technique was adopted to select respondents from three littoral states of the Niger Delta region of Nigeria which were; Akwa Ibom, Bayelsa and Delta States (see figure 1 for details).

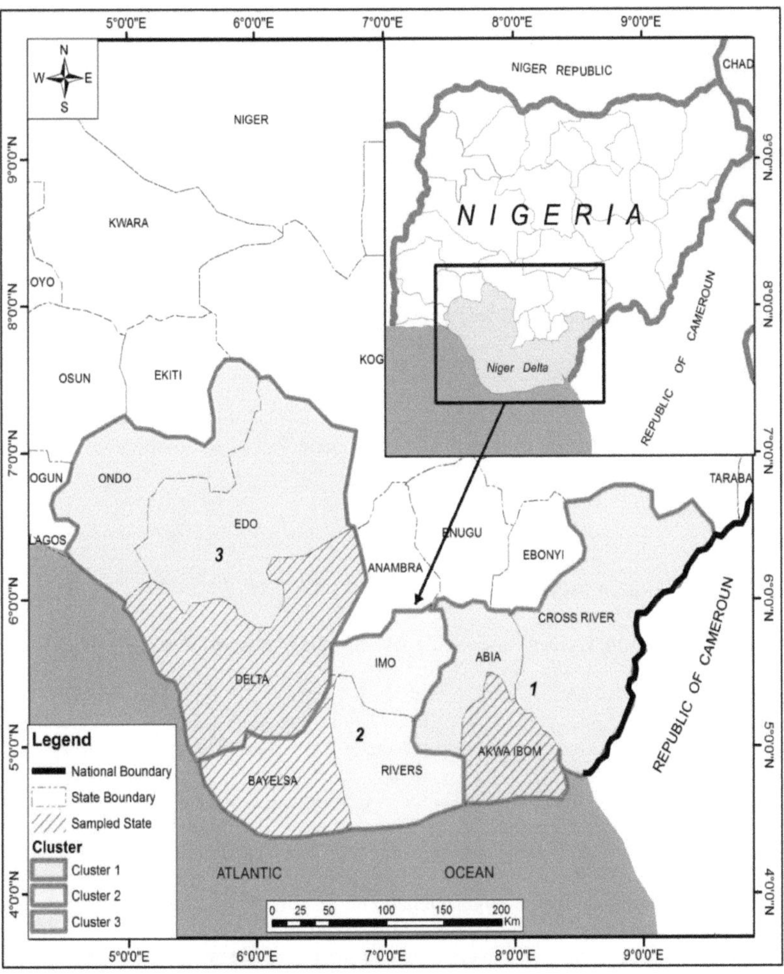

Figure .1: Map of the study Area

Four hundred (400) copies of Likert scale type questionnaire consisting 30 closed-ended questions was designed and administered to the respondents in proportion to the size of each stratum. It is noteworthy that out of the 400 copies of questionnaire that were administered to the respondents, 389 copies were found useable for data analysis as 11 copies were not useful. Thus the study achieved 97.3% questionnaire response rate. Additionally, since it would have been impracticable to conduct interview with all the respondents, selected fishermen, as well as specific market women and speed boat drivers in each of the sample states were interviewed. The respondents answered each statement on the questionnaire and interview schedule based on scales that best described the current situation in their various communities.

The data collected were analysed using descriptive and inferential statistics. Descriptively, simple percentages (%), tables and were adopted for the study, while inferential statistics specifically Pearson Product Moment Correlation (PPMC) was used to test the research hypothesis, in order to make a sound statistical decision. Pearson Product Moment Correlation (PPMC) at 0.05 (r) level of significance was used to determine the relationship existing between the variables being investigated.

Result and Discussion

This section presents results of data analyses on the effect of sea robbery on artisanal fishing in the Niger region of Nigeria measuring.

Sea Robbery and Artisanal Fishing

Research Question I: How does sea robbery affect artisanal fishing in the Niger Delta region of Nigeria?

Artisanal fishing constitutes an important occupation of the people of Nigeria, and the Niger Delta region in particular is noted for its viability in commercial fishing activities. As part of its objectives, therefore this study sought to examine and analyses the effects of sea robbery on fishing business operations in the Niger Delta. The results of the data analysis revealed that sea robbery has affected fishing business operations in diverse ways as summarized in Table 1.1. To be concise, data are presented on two-ordered scale (Agree and Disagree) and were obtained by aggregating the values of "strongly agree and agree" on one hand, and "strongly disagree and disagree" on the other.

Table 1.1: Respondents' opinion on the effect of sea robbery on artisanal fishing categorized by sample groups

Parameters	Fishermen and Women		Speed boat operators		Traders	
	Agree	Disagree	Agree	Disagree	Agree	Disagree
Dwindling means of livelihood	84(98.8)	1(1.2)	57(95.0)	3(5.0)	136(98.5	2(1.5)
Abandonment of fishing business.	68(80.0)	17(20.0)	53(88.4)	7(11.6)	123(72.1)	15(10.8)
Poor fish catch	67(71.6)	18(21.2)	43(71.6)	17(28.4)	121(87.7)	17(12.3)
High cost of fish and other sea foods	75(88.2)	10(11.8)	53(88.4)	7(11.6)	124(89.9)	14(10.1)
Molestation to pay illegal fees	65(76.5)	20(23.5)	34(56.6)	26(43.4)	86(62.3)	52(37.7)

Source: Field Survey (2016).
NB: Figures in parenthesis are percentages

Table 1.1 show a cross-tabulated result on the opinion of respondents (fishermen, speedboat operators, and traders) concerning the effect of sea robbery on artisanal fishing. It is clear that 98.8% of fishermen indicated that their means of livelihood crumbled owing to sea robbery activities, 95% of speed boat operators agreed that they experienced something similar, and the experience of 62.3% of traders was not different. To a considerable extent, sea robbery often lead to abandonment of fishing business as indicated by the opinion of 80% of the fishermen, 88.8% of speed boat operators, and 72.1% of traders. As long as fishermen abandoned their fishing crafts and other fishing gear owing to sea robbery attacks, there is bound to be depletion in fish supply. Clearly 88.2% of fishermen affirmed to low fish yield, 71.6% of speedboat operators had the same view, and 87.7% of traders shared the same opinion.

Owing to fish depletion, poor catch, dwindling income and low level of investment in fishing business resulting from sea robbery activities, respondents were strongly of the view that the persistent high costs of fish and other sea foods in the region could be attributed to robbers at sea. About 88.2% of fishermen, 88.4% of speedboat operators, and 89.9% of traders along the coast of Niger delta waterways supported this view. Finally, most respondents disclosed that they are often forced to make some illegal payments to sea robbers to assuage their predicament. This view cut across 76.5% of fishermen, 56.6% of speedboat operators, as well as 62.3% of traders in the Niger delta region.

Notably, it appears that fishermen along the Niger Delta Seashore are the worst hit by the mayhem of sea robbers. Indeed, statistical evidence (Table 1.1) indicates that fishermen recorded the highest figure in all dimensions of effects of sea robbery activities. For instance, the response for dwindling means of livelihood = 136 (98.5%); abandonment of fishing business = 123 (72.1%); poor fish catch = 121 (87.7%); high cost of fish and other sea foods = 124 (89.9%); and molestation for fees = 86 (62.3%). Qualitative information also corroborates the data collected through the questionnaire instrument. One of the respondents said that:

> The problem of persistent sea robbery has tremendous negative effects on artisanal fishing business. Such effect include poor catch, declining income and low level of investment in fishing business resulting in persistent high costs of fish and other sea foods in the region.

In the views of the respondents, the activity of sea robbers has not only crumbled their businesses, but has chased, robbed them of their means of livelihood. One of the respondents submitted that:

> Owing to the frequent seizure of our engines, lack of money to replace our stolen engines and the risk associated with doing fishing business these days, many of us have abandoned fishing altogether. Right now, some of my friends whose engines were forcefully collected at gunpoint and who cannot replace the engine are out of business. For example, a brand new 75 horsepower Yamaha outboard engine is sold at about 1.3million in the market, while fairly used one cost #700, 000 thousand naira. So, where is the money to replace the ones stolen by sea robbers?

A respondent who confided in the researcher said that, it is risky to do business on the waterways these days because of the activities of sea criminals in Ijaw-kiri waterway. Apart from outright robbery, fishermen are sometimes forced to pay illegal levies to sea robbers and one dare not raise their voice. The so-called monthly dues depend largely on the size of one's boat. The charges range from #7,000 for small boats, while big boats are forced to pay #15, 000. It is compulsory to renew one's dues on the last day of every month as any failure attracts attack.

This situation was confirmed by one of the respondents during an interview session in Oron on 25/4/2016. He said:

> In recent times we are forced to pay dues to sea robbers to guarantee our safety at sea. It is compulsory for every boat to pay #10,000 monthly to sea robbers along Oron/Calabar waterways. The robbers have a record of all the boats that ply the waterways. Sea robbers seized my engine on 20/4/16 along Oron waterways in a broad day light and asked me to put a call across to my people to bring the sum of #50,000. Fortunately, my family was able to raise the money and my engine was release immediately after the payment.

A director of fishing business at Iwuo-Okpom beach in Ibeno shared the same opinion with the above respondent:

> On the last Sunday of a certain month, we went fishing but did not make any catch that day. The next day, I was attacked by sea robbers who hijacked my engine and demanded for #100,000, a roll of Indian hem which costs (#5,000) one bottle of 501 drink which costs (#2,500) and #2000 recharge card. The robbers gave me a phone number to call when coming to pay the ransom for my engine and instructed me not to come with more than two people in the boat.

From the above position, it is evident that the activity of the sea robbers tends to have a significant negative impact on the lives of maritime business operators. As observed during the fieldwork, most maritime business operators affected, who was breadwinners of their families, became handicapped due to the loss of their means of livelihood to sea robbers in the course of their routine business. In an attempt to save their lives, most of these hitherto maritime business operators have abandoned their businesses.

In another interview, one of the respondents from Akwa Ibom State on 28/4/2016 noted that:

> Most fishermen have abandoned fishing business due to regular attacks by sea robbers. I was among the promising directors at Esin-Ofot beach until I was attacked by sea robbers. They carted away all my catch, 200 horsepower Yamaha engine and the fibre, including foodstuffs and cooking utensils and I was abandoned with my boys in the mangrove forest. As a result, I am now handicapped and depend on family members for survival. My 200 horsepower engine, though fairly used was bought with contribution money. Where will I raise money to replace my engine and the boat?

As a result of this incessant maritime criminality, the level of commitment in terms of fishing has significantly depleted such that fishermen tend to restrict their fishing business to shallow waters for security reasons. The effect of this, has been observed in poor fish catch. This is likely to impact negatively on deep sea fishing business.

During the interview conducted on 26/3/2016 in Bayelsa state, one of the respondents lamented that:

> The continuous attacks on business operator in the waterways is life threatening. As such, we limit our activities to shallow waters with very poor catch that is not enough to feed ourselves and our family members. If you happen to fish with fibre boat, the robbers will seize the engine and fibre and abandon you in the nearest mangrove forest. If you are using a local boat, the robbers will steal only the engine and allow the water to carry you to any direction of the tide. Most times, the robbers come to our camps and demand for money. If it coincides with when you just returned from the market and have spent all the money on you, they will seize all your foodstuffs including fuel and then take with them either or destroy any fish they find on the alter or the ones reserved for use.

Based on the results so far analyzed, it can be inferred that sea robbery has had a devastating effect on the fishing business operation in the Niger delta. However, the need to further confirm this assertion calls for testing the hypothesis that no significant relationship exists between see robbery and artisanal fishing as shown on Table 4.4

Test of Hypothesis I

There is no significant relationship between the activities of sea robbers and depletion of artisanal fishing in Niger Delta region of Nigeria.

Pearson correlation analysis was applied on the relevant data to test whether a significant relationship exists between the activities of sea robbers and artisanal fishing. Correlation statistical analysis is an inferential statistics that estimates the magnitude and direction of relationship between two or more variables. The correlation coefficient (r) ranges from -1 to +1 and the closer the r-value to +1, the stronger the relationship between the variables under study.

Table 2.2: Pearson's correlation matrix showing relationship between sea robbery and
artisanal fishing

		Sea Robbery Activity	Artisanal Fishing
Sea Robbery Activity	Pearson Correlation	1	.521[**]
	Sig. (2-tailed)		.000
	N	389	389
Artisanal Fishing	Pearson Correlation	.521[**]	1
	Sig. (2-tailed)	.000	
	N	389	389

**Correlation is significant at the 0.01 level (2-tailed).

Result of the statistical analysis in Table 2.2 indicates that there is a significant relationship between sea robbery and artisanal fishing (r=0.612; p<0.01) which necessitates to the rejection of the null hypothesis at 0.05 level of significance. Given the strong negative correlation of 0.612, this implies that a percentage change (increase or decrease) in sea robbery activities would lead to a corresponding change in artisanal fishing and vice versa. In other words, the more fierce the menace of sea robbery in Nigeria territorial waters, the greater the havoc done to fishing as a business in terms of poor fish catch , substantial loss of income, downturn in means of living, and high cost of available sea foods in the market.

Conclusion/Policy Implications

The main thrust of this study was to examine the impact of sea robbery on artisanal fishing: the Niger Delta region of Nigeria experience. Findings from data analyzed revealed that sea robbery has adverse effect on the fortunes of artisanal fishermen in the Niger Delta region. For instance, fish depletion, which is responsible for poor catch and the high cost of fish and other sea food currently experienced by Nigerian living in littoral states and its attendants consequences such as dwindling means of livelihood, has prompted some fishermen to abandon their lucrative fishing business. These findings gain support from the works of Usim (2016); Ochai (2013); Mpi (2011) among others.

Consequent upon this finding, the study suggests the following recommendations:
(1) Community policing especially in hotspots communities is recommended, this type of policing is result oriented and more effective and efficient compared to the traditional conventional random policing
(2) Fishermen and other maritime business operators are advised to operate cashless, judging the risk involved in carrying money along the creeks and waterways. Maritime business operator should devise other method of exchange if possible trade by barter.

REFERENCES

Aderigbola, D. O. (2015). Emergence of the Gulf of Guinea in the Global Economy: Prospects and Challenges. IMF Working Paper: Office of the Executive Director- Africa, WP/05/235

Cohen, L. E. and Felson, M. (1979). Social Change and Crime Rate Trends: A Routine Activity Approach. *American Sociological Review*, 44, 588–608.

Cullen, F. T. and Agnew, R. (2006). *Criminological Theory, Past to Present: Essential Readings* (3rd Edition). Los Angeles: Roxbury Publishing Company, p. 446.

Dogarawa, L. B. (2013). Sustainable Strategy for Piracy Management in Nigeria. *Journal of Management and Sustainability*, 3(1): 119–128.

Ezem, F. (2012). Challenges of Curbing Criminality in Nigeria's Maritime Domain. *National Mirror*, December 28, p. 5.

Ezem, F. (2013). Worsening Sea Piracy Threatening Foreign Direct Investment Inflows. Available at: http://nationalmirroronline.net/new/worsening-sea-piracy-threatening-foreign-direct-investment-inflows/. Accessed on March 2, 2015.

George, R. (2015). Pirates Step up Attacks on Vessel in Gulf of Aden, Off Somali Coast. Available at: http://www.voanews.com/a/pirates-break-southeast-asis-busy-gulf-guinea/3280786.html. Accessed on March 10, 2016.

Graf, A. (2011). Countering Piracy and Maritime Terrorism in South East Asia and off the Horn of Africa – Applying the Lessons Learned from the Countermeasures against Maritime Violence in the Straits of Malacca to the Gulf of Aden. A PiraT-Working Paper on Maritime Security, 5, 1–58.

Igbo, E. M. (2008). *Aetiology of Crime: Perspectives in Theoretical Criminology*. Enugu: New Generation Books, p. 180.

International Chamber of Commerce (ICC) (2017). Piracy and Armed Robbery against Ships. 2013 Annual Report. London: International Maritime Bureau. Available at: www.icc.ccs.org. Accessed on February 24, 2017.

International Maritime Bureau – IMB (2009). *Piracy and Armed Robbery against Ships: Report for the Period January, 2008 – December, 2009.* London: International Chamber of Commerce, p. 98.

International Maritime Organization (IMO). (2014). Code of Practices for the Investigation of Crimes of Piracy and Armed Robbery against Ships, International Maritime Organization Assembly Res. A 1025(26), Annex, 2 January Annual Report.

Jamabo, N. A. and Ibim, A. T. (2010). Utilisation and Protection of the Brackish Water Ecosystem of the Niger Delta for Sustainable Fisheries Development. *World Journal of Fish and Marine Sciences*, 2(2): 138–141.

James, l. (2013). Maritime Strategy in Globalizing World. *Orbis,* 51(4): 569 –575.

Morrow, J. (2015). Routine Activity Theory. Available at: http://criminology. wikia.com/wiki/Routine_Activity_Theory. Accessed on May 22, 2015.

Mpi, K. (2011). Unemployment, Cause of Sea Piracy. Available at: http://www.thetide newsonline.com/2011/02/18/%E2%80%98unemployment-cause-of-sea-piracy %E2%80%99/. Accessed on March 2, 2015.

Nicoll, G. (2012). Combating Piracy and Oil Theft in Nigeria. Available at: http://www.lookoutnewspaper.com/combating-piracy-and-oil-theft-in-nigeria/. Accessed on February 24, 2015.

Nurudeen, N. A. (2015). Nigeria: Sea Robbery Impediment to Local Fish Production. Available at: http://allafrica.com/stories/201405150590.html. Accessed on February 3, 2015.

Ochai, P. (2013). Owners of Fishing Trawlers Want the Federal Government to Save their Business from the Increasing Attacks they Encounter from Sea Pirates. *Thisday Newspapers*, January 7, p. 2.

Ojo, A. Y. (2016). Africa's Maritime Dimension: Unlocking and Securing the Potentials of its Seas – Interventions and Opportunities. *African Security Review,* 16(2): 112 - 124.

Okhomina, O. (2016). Piracy: Armed Attacks from Foreign Fishing Trawlers Spark Violent Protest in Bayelsa. Available at: http://leadership.ng/news/497734/ Piracyarmed-attacks-foreign-fishing-trawlers-spark-iolent-protest-bayelsa. Accessed on March 8, 2016.

Onuoha, F. C. and Hassan, H. I. (2009). National Security Implications of Sea Piracy in Nigeria's Territorial Waters. *The Nigerian Army quarterly Journal,* 5(1): 1 – 28.

Orji, U. J. (2013). Tackling Piracy and other Illegal Activities in Nigerian Waters. *Journal of Defense Resources Management*, 2, 65 – 70.

Otto, L. (2015). Maritime Crime in Nigeria and Waters beyond Analysing the Period 2009 to 2013. *Africa Journal online*, 45(1): 15 – 35.

Oyetunji, A. (2012). Nigeria Loses 2 Trillion Yearly to Oil Theft, Piracy. *The Nation Newspapers*, July 23, p. 4.

Perouse de Montclos, M. A. (2012). Maritime Piracy in Nigeria: Old Wine in New Bottles? *Studies in Conflict and Terrorism,* 35, 531 – 541.

Phelps, S. (2010). Nigeria Maritime Security, the Reality. Available at: www.marsecr eview.com/2010/nigeria-maritime-security-thereality. Accessed on October 5, 2015.

Pittman, B. (2011). Economic Impact of Maritime Piracy. *Africa Economic Brief*, 2(10): 1 – 8.

Sanga, J. M. (2010). Countering Persistent Contemporary Sea Piracy: Expanding Jurisdictional Regimes. *American University Law Review*, *59*, 1267–1319.

Ships and Ports. (2014). Fishermen Lament Losses to Sea Robbers in Akwa Ibom. Available at: http://shipsandports.com.ng/fishermen-lament-losses-to-sea-robbe rs-in-akwa-ibom/. Accessed on May 19, 2015.

Tawari, C. C. and Davies, O. A. (2010). Impact of Multinational Corporations in Fisheries Development and Management in Niger Delta Nigeria. *Agriculture and Biology Journal of North America*, 1(2): 146–151.

Terzi, G. (2012). Maritime Piracy: A Threat to our Security and the Global Economy. Available at: http://www.esteri.it/mae/en/sala_stampa/archivionotizie/interviste /2012/10/20121004_marpirindonesia.html. Accessed on April 10, 2015.

Udensi, L. O., Okpara, E. N. and Oyinyechi, C. E. (2014). National Security and Maritime Piracy in Nigeria: Sociological Discourse. *Humanities and Social Science Letters*, *2*(1): 60–71.

United Nations. (1982). United Nations Convention on the Law of the Sea. Available at: http//:www.imo. org/facilitation/mainframe.asp?topic_id=362. Accessed on August 29, 2015.

United States Legal Incorporation. (2015). Routine Activities Theory. Available at: http://definitions.uslegal.com/r/routine-activities-theory-criminology/. Accessed on May 22, 2015.

Usim, L. (2014). Nigerian Sea Pirates Free Tanker after Ransom. *Leadership,* January 11, p. 12.

Usim, U. (2016). Sea Pirates Killing Nigeria's Fishing Industry. Available at: www.sunnewsonline.com/new/sea-pirates-killing-nigerias-fishing-industry. Accessed on January 10, 2016.

Wajilda, J. A. (2013). An Overview of the Economic Implications of Piracy and Armed Robbery against Ships in Nigeria. *WMU Studies in Maritime Affairs*, *2*, 125–135.

Wikstromolof, H. (2009). Routine Activity Theories in Criminology: Oxford Bibliographies. Available at: http://www.Oxfordbibliographies.com/view/ document/obo-9780195396607/obo-9780195396607-0010.xml. Accessed on May 22, 2015.

Yamame, T. (1967). *Statistics: An Introductory Analysis* (2nd Edition). New York: Harper and
Row, p. 345.

Zircon Marine Ltd., (2014). Sea Robbers Attack Fishermen, Snatch 18 Outboard Engines.
Available at: http://zirconmarine.com/sea-robbers-attack-fishermen-snatch-18-
outboard-engines/. Accessed on May 18, 2015.